THE GREAT
BRUNEL

A PHOTOGRAPHIC JOURNEY

Chris Morris

Introduction by

Neil Cossons

AMBERLEY

First published 2005 by Tanners Yard Press
Third edition by Amberley Publishing, 2015

Amberley Publishing
The Hill, Stroud, Gloucestershire, GL5 4EP
www.amberley-books.com

Designed by Paul Manning

British Library Cataloguing in Publication Data
A catalogue record for this book is available from the British Library

ISBN 978 1 4456 5079 1 (Print)
ISBN 978 1 4456 5080 7 (Ebook)

The author wishes to thank the following
for permission to reproduce copyright material:

City & County of Swansea Museum Service Collection:
image of Landore Viaduct on page 92.

National Museum of Science and Industry Library:
images from 'Steam' museum, Swindon, pages 9, 47, 44 and 76.

CONTENTS

Clifton Suspension Bridge.

Preface to the third edition

While working on this project I was asked: 'Another book on Brunel – what is your angle?' That was easy for me: because I am a photographer this is primarily a visual celebration of the great man's achievements and, as far as I know, no comparable collection of photographs exists in print.

The title of the book reflects both the esteem in which Brunel is held today and the certainty with which he named his own projects (the 'Great' Western Railway and the three 'Great' steamships); the subtitle 'A Photographic Journey' supports my contention that industrial history can be appreciated by visiting old artefacts and sites as well as in a library.

Believing that the concept of the original designs remains intact, I have not tried to obscure the many alterations to Brunel's engineering; neither have I been shy of including the everyday trappings of the twenty-first century: high-visibility waistcoats, diesel trains, tarmac and traffic lights all happily interact with the old legacy to provide a vibrant living history. Some photographs are included which are not of Brunel's work but help to illustrate his story; some of the railway photographs show typical design patterns that may not be original.

The contents do not run strictly chronologically as Brunel worked on many schemes at once; the railway chapters run mainly geographically. The commentary is informal but should provide sufficient technical information to allow readers to appreciate the images. Although in general there are no reproductions of plans or old photos, I did find a way, at the end of the book, of including Robert Howlett's famous photograph from Millwall.

While a few Brunel sites are very well known, a surprising number are literally unnoticed; luckily I was able to draw on the enthusiasm of the many organisations and individuals who helped me locate and photograph these items (a list of acknowledgements appears at the back of the book). I am particularly grateful to Steven Brindle for checking my text, and to Sir Neil Cossons for drawing on his expertise and experience to provide the introduction that follows.

The Great Brunel has been unavailable for some time. I am very grateful to Amberley Publishing for keeping the book in print with this third edition.

Chris Morris, 2015

INTRODUCTION

Isambard Kingdom Brunel was a revolutionary born into a world of revolution. He grew to be a hero in an age of heroes. By any measure one of Victorian Britain's greatest engineers, what set Brunel apart from his contemporaries was the scale of his ambition, the span of his intellect, and a capacity to excite controversy through his ideas and notions. This contentiousness was borne not of perversity so much as supreme ability, and an unparalleled self-belief reinforced by a powerful and effervescent personality. Brunel was a polymath whose genius lay in his spontaneous and eclectic response to challenge. His works are his monument, unequalled in range and diversity. His successes were audacious in scale, breathtaking in their daring and panache. Even his failures were larger than life. Brunel was not a man for half measures. Through Chris Morris's outstanding photographs, this book seeks to evoke something of that spirit.

On 14 September 1805 Vice Admiral Horatio Nelson boarded HMS *Victory*, for the last time from British soil, and sailed for Cadiz. Five weeks later he was dead but the combined French and Spanish fleets had suffered a defeat so decisive that British supremacy at sea was unchallenged for more than a century. On the afternoon of his departure Nelson inspected the newly completed Blockmills at Portsmouth, where he saw the automatic machinery – among the first mass production equipment in the world – supplying the pulley blocks for the rigging of its ships that the Royal Navy so desperately needed. His expressions of pleasure at what he saw might well have offered encouragement to their designer, Marc Isambard Brunel, then locked in dispute with the Admiralty over payment for his work. But what is perhaps more certain on that September day is Brunel's delight that his wife Sophia was newly pregnant with their third child who was to be born, in their modest house in Portsea, on 9 April of the following year. This was Isambard Kingdom Brunel.

The age into which the younger Brunel was born was by any measure an auspicious one. As the post-war economic depression following the defeat of Napoleon lifted, Britain entered a period of unprecedented growth to emerge by the middle years of the nineteenth century unchallenged as the world's leading power, a supremacy borne of mercantile, industrial and imperial influence.

Growing up in the intellectual ferment of this new world was to provide the young IKB with opportunity beyond compare. Educated at his father's knee, in France, and then as an assistant on the elder Brunel's own extraordinary engineering works, his early experiences gave him a grounding without equal. By 1827, at the age of twenty-one, he was Resident Engineer on the Thames Tunnel, the engineering project which was to test not only his abilities as an engineer but qualities of leadership and determination too. The young Brunel rose to the challenge. His experiences under the Thames were to stand him in good stead for the rest of his career. Four years later, on 16 March 1831, his design for a suspension bridge across the Avon Gorge at Clifton was accepted and in December of the same year he travelled for the first time by train, on the Liverpool & Manchester Railway, the world's first passenger-carrying steam railway which had

Facing page: Royal Albert Bridge, Saltash.

Above: SS *Great Britain*, Bristol.
Facing page: Disc and crossbar signals, 'Steam' museum, Swindon.

opened some fifteen months earlier. He wrote, 'The time is not far off when we shall be able to take our coffee and write while going noiselessly and smoothly at 45 mph. Let me try.' He did of course try and the Great Western Railway was the result, a line renowned for its speed, comfort and smoothness, not least because of the broad gauge that Brunel had insisted upon.

I cannot remember a time when Brunel was not in one sense or another a part of my life. My father, brought up in Chippenham in the 1890s, had already inured me with the mystique of the Great Western by the time my mother in the 1950s was writing scripts on Brunel for BBC schools broadcasts. As curator of the Great Western Railway Museum at Swindon in the early 1960s I came to see and understand at first hand his genius, in his beautiful wash drawings of Clifton suspension bridge and through John Cooke Bourne's beautiful lithographs of the railway itself, both in the museum's collections, and in the buildings of the railway works and the adjacent housing. By 1964 I was living in Clifton in time for the centenary celebrations in December of that year of the suspension bridge and with Brunel's Temple Meads trainshed and his Bristol harbour works on my doorstep. And I was there when a miscellaneous group of Brunel aficionados met for the first time, in Bristol City Museum, to debate the possibility of returning the *Great Britain* from exile in the Falkland Islands to the city of her birth. In 1985 I was able to see one of her masts preserved on the waterfront in Port Stanley, the last substantive relic of the ship to survive there. And I relished the opportunity to leaf through the Brunel notebooks in the University Library. Here were designs for guns, a gas jet, a bucket, the Great Western Railway, a gunboat, Calcutta station, decorative numbers for railway locomotives, three great ships and seven detailed drawings for an inkstand. Brunel left nothing to chance.

Later, as Chairman of English Heritage – now Historic England – I was able to delight in the rescue of the components of Brunel's Bishops Road Bridge, Paddington for eventual re-erection and to make regular visits to the National Monuments Record, itself housed in the offices of Brunel's Great Western in Swindon.

In this book Chris Morris has explored most of Brunel's extant works to offer an extraordinary visual feast, powerfully expressive of the works – and through them of the character – of this extraordinary man. And, rather in the manner of the Brunel notebooks, he ranges from the grand gesture to the minute detail, reflecting Brunel's obsession with controlling every aspect of those works for which he was responsible.

As a young man Brunel wrote, 'My self-conceit and love of glory or rather approbation vie with each other which shall govern me'. He craved recognition; today he has it. At first eclipsed by the eminence of his father, by the time of his death IKB was seen as a great, if enigmatic, figure whose reputation was perhaps clouded by the recent difficulties associated with the *Great Eastern* steamship. But over the years the qualities of the younger Brunel have matured and blossomed in the public imagination in line with our wider appreciation of the achievements of the Victorian age. Chris Morris's illustrations give us the opportunity to forge a new perspective within which to celebrate the heroic nature of the man and his works.

Neil Cossons

London Borough of Southwark

Isambard Kingdom Brunel
1806-1859

Great Victorian Engineer

His first project was the Thames Tunnel, the world's first underwater tunnel

Voted by the People

1

BEGINNINGS

Brunel's father, Marc Isambard Brunel, came from a prosperous farming family in Normandy. A declared Royalist during the Revolution, he was forced to flee to America, where he spent six years working as an architect and engineer. In 1799 he sailed for England, having perfected an invention which would facilitate the manufacture of 'blocks' used in ships' rigging which he hoped to sell to the British navy. He was also more than keen to be reunited with Sophia Kingdom, whom he had met in France and who was then living in London.

Marc Brunel married Sophia later the same year; they lived in Portsea, close to Portsmouth naval dockyard. Preceded by two sisters, Isambard Kingdom was born in April 1806. His education was to be largely in France and, crucially, as his father's apprentice and assistant.

Marc Brunel pursued an inventive but precarious life (including a spell in debtors' prison caused by a cancelled contract and late payment by the Admiralty). In 1818 he patented his most far-reaching concept: a 'shield' for tunnelling made from huge sections of cast iron, which protected the tunnellers from falls while the newly excavated sections were bricked up. All attempts to tunnel under the Thames had failed, but a new act of parliament allowed the Thames Tunnel Company to try again in 1825, and Marc Brunel was appointed engineer.

Rapid progress was made, but within a year Marc Brunel became very ill and Isambard, at the age of twenty, took on the job of engineer in charge. In 1827 a shingly section of the river bed collapsed into the tunnel; personal bravery by the young Brunel, descending by rope into the flood, saved the life of a tunneller. Undaunted, the Brunels supervised a restoration of the unfinished work, only to suffer more flooding months later. This time the young Brunel was working at the tunnel face and had a lucky escape: as the alarm was raised, his inert body was washed up to the surface in the shaft. Six men who had been working with him died. Work on the tunnel was suspended and Brunel was sent to Clifton to convalesce.

After the tunnel disasters Brunel felt frustrated and unfulfilled, looking on as his contemporaries were appointed to big projects while he could only dream. Then in 1830 he won a competition to build a bridge over the Avon Gorge at Clifton: his career was about to take off, and Bristol was to be the epicentre of his world.

The merchants of Britain's premier west coast port had watched ruefully as Liverpool, the upstart, eroded their trade. Already Brunel had made improvements to Bristol harbour, but his further advice – to abandon the narrow tidal Avon and site a new port some ten miles away on the Severn – had been rejected.

By 1833 a group of city worthies had a new scheme to defend their position: to join the modern world with a railway from Bristol to London. Brunel (enrolling as a special constable at the time of the 1831 riots) had made himself a Bristolian, and it was to him that Thomas Guppy, leader of the railway committee, turned for technical help.

Facing page: Blue plaque at Rotherhithe.

The Thames Tunnel

Of all the projects undertaken by Brunel's inventive and prolific father, the Thames Tunnel is the one for which he is best remembered. Its tortuous history began in 1825 with the sinking of a shaft at Rotherhithe. In 1827 the young Brunel, only twenty years old, became resident engineer. After two floods, work on the tunnel was suspended. By the time work resumed in 1835, Brunel was busy on his own account and was not further involved: the tunnel finally opened in 1843.

The pump house (*right and far right*) still stands by the river in Rotherhithe; adjacent is the shaft which drops down to the tunnel (*top*), today used by the East London underground line.

Facing page: On the north side of the river the twin tube tunnels emerge at Wapping station on the East London overground line.

Brunel's Bristol

Sent to Clifton to recuperate after the Thames Tunnel accident, Brunel rapidly made Bristol his home from home. He won a competition to design a bridge over the Clifton Gorge (*facing page and page 22*), which was the beginning of a long association with the city.

The dock where the SS *Great Western* was built is now the site of the Bristol Industrial Museum, and Brunel's statue (*right*) stands on Broad Quay nearby. Further down the harbour lies the SS *Great Britain* (*pages 102, 104*). Just to the south of Broad Quay is Queens Square, where Brunel enrolled as a special constable at the time of the Bristol riots in 1831. To the west is Brunel House (*top right*), formerly the Great Western Hotel, where rail passengers from London could stay overnight before embarking on their transatlantic voyage.

Below: The clock with two minute hands, on the front of the Exchange in Corn Street, told both London and local time – a feature of provincial life that was ended by the advent of train timetables. Further to the east of the city centre is Temple Meads Station (*page 26*).

S.S. "GREAT WESTERN"
(DESIGNED BY I.K. BRUNEL)
THE FIRST STEAMSHIP BUILT FOR & ENGAGED IN
REGULAR TRANSATLANTIC TRADE WAS LAUNCHED
NEAR THIS SPOT ON THE 19TH OF JULY 1837
LENGTH 236 FEET. BEAM 35 FEET
FIRST VOYAGE
BRISTOL TO NEW YORK
8TH TO 23RD APRIL 1838.

Facing page: The elegant curving masonry of Brunel's Cumberland Basin south lock was also designed to help clear silt. It is now out of use but remains as built. Adjacent is the original swing bridge (*left*), a riveted tubular design which led to his famous work at Chepstow and Saltash.

At the start of the nineteenth century Bristol's docks consisted of wharves on the banks of the Avon, where ships canted into the mud at low tide. William Jessop addressed the issue by diverting the river and putting gates across its old course, thus creating the 'Floating Harbour'.

To help clear away the silt which was constantly being washed into the harbour by the Avon, Brunel advocated replacing a weir that controlled the high water level of the dock with an 'underfall' – a low sluice to help flush away the mud. 'Scouring' is used to this day to keep the channels free (*left*).

Bridgwater Dredger

Continuing concern over the silting up of Bristol's floating harbour led Brunel to invent a dredger. Powered by an onboard steam engine, the vessel hauled itself along on a chain (*right and far right*) which was fixed between points on the quayside. A blade rather like that of a bulldozer was lowered into the mud to scrape it into the channel, where it could be flushed away.

Facing page: The dredger used in Bristol is in pieces in the industrial section of Bristol's 'M Shed' museum. These photographs are of a similar one made for Bridgwater docks and can be seen as part of the 'World of Boats' museum in Eyemouth, north of Berwick.

Docks and Drains

Despite the improvements in his career prospects following his move to Bristol, Brunel was not shy of taking on humble commissions.

Some stonework remains at Monkwearmouth (*facing page*) where Brunel rebuilt the dock in 1831. The Fossdyke (*right*), a navigation since Roman times from the River Trent at Torksey Lock (*below*) to Lincoln, was improved by Brunel in 1833.

Clifton Suspension Bridge

While recuperating in the Clifton
district of Bristol, Brunel entered
the competition to design a
bridge over the Avon Gorge.
All the entries were vetoed
by an ageing Thomas Telford
so a re-run was held in which
Brunel came second. It tells
us of Brunel's character that,
undaunted, he met the judges and
persuaded them to declare him
the winner.

The eastern suspension tower of
the bridge is clad in scaffolding
during long-term conservation.

Clifton Suspension Bridge

After the abutments were built, a 600-foot iron rod made up of welded lengths was suspended across the gorge (a section is shown by Andy King, *far right*, senior curator of the industrial section of Bristol's 'M Shed' museum). Brunel was the first to cross, in a basket suspended on a pulley wheel pulled by rope: when it got stuck in the middle, he climbed up to free the snag!

Funds for building the bridge ran out in 1842 and work was aborted. It was eventually finished, to a modified design, as a memorial after Brunel's death. The suspension chains, which had been sold for use on the Saltash bridge, were replaced with those from Brunel's dismantled Hungerford Footbridge.

2
BRISTOL RAILWAY

In 1833 most of the country's rapidly spreading railway network was devoted to carrying commercial goods. Almost all the lines were in the industrial north, and, with the exception of a few old 'tramways', all were built to a standard gauge of four feet eight-and-a-half inches. The proposal to join London to Bristol would create a single railway far longer than anything in existence. The railway's sponsors saw a market in carrying passengers as well as freight traffic to London from the wealthy counties to the west. Their newly appointed engineer, Brunel, who had done no such work before, decided the rail gauge should be seven feet and a quarter of an inch – though prudently he did not reveal this to the directors until the bill was safely through parliament in 1835.

Brunel's argument was that a broader gauge made possible a smoother, more stable and potentially faster service. His only precedent was that his father had laid a wide gauge track in Chatham docks for transporting timber. Instead of the conventional railway track of rails fixed on sleepers set in ballast, Brunel's broad gauge utilised heavy timbers laid under the track longitudinally, with cross ties for spacing every fifteen feet. This was the exact layout of the Chatham tracks. At the time no one considered how connections would be made with other railways – a fact which was to cause enormous difficulties later.

The route Brunel chose has been universally acknowledged the best. From London to Reading it followed the Thames, but where Brunel might have been expected to follow its tributary, the Kennet, towards Newbury (as did the Kennet and Avon canal and the ancient Bath road route), he stayed with the big river, through the gap in the hills at Goring. Finally leaving the Thames near Didcot, the railway took a level route through the Vale of the White Horse, a gallop of a line towards Swindon and Chippenham. Did Brunel have the lucrative Oxford connection in mind? Or was he already spreading the broad gauge in as wide a wedge as possible, staking out territory towards the Midlands in order to bolster his maverick choice?

The line from London to Swindon, so smooth and level that it was known as 'the billiard table route', finally met the steep edges of the Cotswolds approaching Bath. The proposed tunnel under Box Hill would be almost two miles long, far greater than anything previously attempted. The general scepticism of Brunel's opponents and cynics was compounded by the relatively steep gradient of 1 in 100 throughout the tunnel. His detractors' dire warnings of runaway trains were to prove unfounded.

In constructing the railway, Brunel's contractors relied heavily on the canal network. Rail and other materials were delivered to Bulls Bridge on the Grand Union, adjacent to the line in Hanwell, west of London; further along the line the Wiltshire and Berkshire Canal and the Kennet and Avon canals were similarly useful. Assisting the railways was like writing a suicide note and as the lines opened, the canals went into rapid decline.

Facing page: Disc and crossbar signals, Didcot Railway Centre.

The original Paddington of 1840, situated to the west of the present station, was of a temporary nature. Wanting a grand terminus to rival Euston, in 1851 Brunel persuaded his financial masters of the need for a new high-profile London presence.

The glass and iron roof (*right*) was inspired by his friend Joseph Paxton's work on the Great Exhibition (later Crystal Palace), a project on which Brunel had served as a committee member.

Details such as the 'Director's Balcony' (*above*) were provided by Brunel's architect associate, Digby Wyatt, while the chateau-style hotel fronting the station on Praed Street (*top left*) was designed by P. C. Hardwick.

5
Welcome
to
Paddington
Station
This station
is owned
and
managed by
Network Rail
Welcome
to
Paddington
Station
This station
is owned
and
managed by
Network Rail

Bishops Bridge

In 2003, during demolition work on Bishops Bridge (immediately to the west of Paddington), the remains of an original Brunel bridge were found buried in the later structure. Brunel had carried Bishops Bridge Road across his tracks on masonry arches, but had used cast iron for the two sections which crossed the adjoining canal. In 1906 the brick arches were replaced with riveted steel, and the iron bridge was hidden under masonry additions.

The structure is of wrought iron tubing resembling that preserved at Bristol's Cumberland Basin, but pre-dating it. The dismantled bridge (*facing page*, in a snowstorm) with the ironfounders' name 'Gordons' cast into some sections (*right*) was removed to a store at an English Heritage depot in Portsmouth. Twelve years on, the plan to reassemble the bridge to cross the canal close to its original location has made no progress.

There is a common misconception, fuelled by accounts in the press, that the bridge hanging melodramatically over the ongoing engineering works (*above*) is Brunel's. In fact, this is the 1906 steel structure, awaiting an easier and less disruptive demolition when a new bridge is built beneath it.

Wharncliffe

At Hanwell the lines of the Great Western cross the River Brent on an eight-arch brick viaduct. The coat of arms of Lord Wharncliffe, the chief sponsor of the Great Western Railway bill in the House of Lords in 1835, is prominently displayed as an acknowledgement of his support (*facing page*).

Brunel's fascination with Egyptian style, already manifested in his Clifton drawings, reappears in the design for the capitals on the double line of columns (*above and right*). The original section of the viaduct is the south side, subsequent widening having added a third line of piers to the north.

Canals played a prominent role in the construction of the railways. The line of the Great Western crosses the Grand Union at Bulls Bridge Wharf (*facing page*), which Brunel used to deliver materials. Also in Hanwell is Windmill Bridge (*right*), a triple-layer crossing where road and canal fly over the Brentford branch railway.

Brunel also made use of the Kennet & Avon for the western lines at Bath and the Wiltshire and Berkshire through the Vale of the White Horse; Dauntsey Wharf, near Chippenham (*far right*) had a direct connection to the railway.

The famously flat arches of Maidenhead Bridge (pages 38-9) had their profile dictated by the Thames Commissioners, who demanded a wide passage for boats, and by Brunel's reluctance to make the railway gain height for the crossing. A more prosaic bridge (*above right*) carries the railway over the River Kennet at Reading.

Although the river valley provided a level route there were certain obstacles, exemplified by the trains deep in the gloom of the cutting through Sonning Hill (*facing page*). With one assistant to help him from the Bristol end, and another, Hughes, for the eastern parts of the line, Brunel worked exhausting hours on the survey. His team may not have shared his work ethic: 'looked for Hughes everywhere and found him at breakfast in the Black Boy' (pub sign, *above left*) notes the boss scathingly.

A Branch to Oxford

At Gatehampton Bridge (p. 124) boats go straight on towards Oxford while trains finally abandon the Thames valley, passing through Didcot to reach Steventon at the beginning of the Vale of the White Horse. An important station, Steventon provided a temporary terminus and board meetings were held in the Superintendent's House (*right*) during 1842–3.

A branch line from Didcot to Oxford was first planned in 1833 but was held up by opposition from the university. From 1840 Steventon had acted as a stopgap, with eight stagecoaches a day making the ten-mile connection to the city.

Facing page: a Network Rail gardener strims weeds at Culham station on the Oxford branch line, which finally opened in 1844.

Gooch's Fliers

The original specifications for locomotives which Brunel put out to manufacturers were very restrictive: the stipulated cylinder size and piston speed made it impossible to attain the performance expected. Luckily in 1837 the company appointed Daniel Gooch, not yet twenty-one years old, as 'locomotive assistant'. Brunel had already ordered two engines, *North Star*

and *Morning Star*, from Robert Stephenson, Gooch's previous employer. Gooch modified Stephenson's design to create *Firefly*; a production run of 62 locomotives followed, providing the basic motive power for the Great Western for two decades.

North Star was in service until 1870; it was stored in Swindon but in 1906 was deemed to be taking up too much space and was scrapped. A replica is in 'Steam' museum at Swindon (*right*), as is the marble bust of Daniel Gooch (*left*).

Facing page: a modern working Firefly, number 63, was built at Didcot Railway Centre for Brunel's bicentenary in 2006.

In 1840 Brunel decreed that locomotive staff should wear white suits but five years later sensibly allowed blue as an alternative. On the footplate of the new engine (still awaiting its brass steam valve cover), Sam Bee, chairman of Firefly Trust, relives the original dictum on a test run at Didcot in June 2005.

Swindon

It was Gooch who chose Swindon as the location of the GWR's engineering base. Swindon was less than halfway from Bristol to London, but Gooch, still thinking a little like a stagecoach operator, felt it to be well positioned for a refreshment break and a change of engine. (It was Gooch's belief that two different types of locomotive would be used on the Great Western: one to tackle the grades from Bristol to Swindon and another to race the level line to London.)

For a long time the refreshment stop hindered the development of a high-speed train service: the dining rooms were let out long-term to a caterer on the understanding that all trains would stop for about ten minutes – a condition to which the lessee rigorously clung. Brunel once famously complained about the quality of the catering, but his vitriolic remarks about the coffee may also have been fuelled by resentment about the deal.

Formerly a tiny village, Swindon mushroomed into a railway town, with Brunel himself designing the Mechanic's Institute (*above*) and the first houses (*right*). A huge complex of buildings sprang up, forming the largest railway construction and maintenance workshop in the world. 'Steam' museum today is housed in part of the original Swindon buildings. The Swindon workshops were constructed from stone excavated from Box tunnel (page 50).

Facing page: Brunel's folding surveyor's stick is in 'Steam' museum, Swindon, displayed here by collections officer Elaine Arthurs. Brass protrusions on the stick define the measurements of broad gauge and – surprisingly considering how defiantly Brunel defended his choice of track width – that of standard gauge too.

Chippenham

Chippenham also benefited from the Box Tunnel stone, though its viaduct (*right*) is now repaired with engineering bricks. The small building in front of the station (*facing page*) was used as an office by Brunel.

Box Tunnel

The tunnel through Box Hill was the longest ever built. It was dug by a team of 4,000 navvies, and there were 100 fatalities. Its gradient of 1:100, dropping down towards the Avon valley, led pessimists to predict disastrous runaway train scenarios. The doomsters were proved wrong, and in 1844 the first locomotive completed the London to Bristol route without mishap.

Right: the western portal of the tunnel exemplifies the elegance and care Brunel applied to even his most humble projects. Vast quantities of honey-coloured Bath stone provided a glorious material for bridges and buildings along the line.

Facing page: Unusually, for most of its length the tunnel is not lined with brick but remains as a huge linear cavern. The view here is of a Network Rail maintenance crew, at the point where the soot-stained limestone finally gives way to a brick lining to the eastern entrance.

50

Facing page: The Great Western Railway enters Bath from the east through Sydney Gardens. The gloom of the high retaining wall to the south is mitigated by the low balustrading edging the park, where benches provide an unusually close-up view of the Great Western. Brunel had a distrust of cast iron under stress but presumably thought it adequate for the footbridge (which, although dated 1865 in the *Institute of Civil Engineers' Handbook*, appears in a contemporary lithograph of the track being laid).

The price paid for keeping the line away from Georgian Bath is that the southern parts of the town are dominated and divided by long stretches of viaduct and high-walled embankment, as at the eastern Dolemeads Viaduct (*right*).

Opposition to the coming of the railway to Bath posed a problem for Brunel. His great and lasting achievement was to bring the line into a central station without impinging on the historic areas of the town. The tracks stay south of the Avon, crossing and re-crossing (where the river takes a large meander) just to provide for the station. St James's Bridge (*above*) is heavily repaired using iron tie-bars; sections of Bath stone have been replaced with more durable engineering bricks.

Bristol Temple Meads

Although London had to wait until 1851 for its grand terminus at Paddington, the financial masters at the Bristol end of the line had allowed Brunel more leeway: Temple Meads, which opened in 1840 at the same time as the railway itself, was a stylish concept. Progressive changes from 1841 onwards to accommodate the Exeter lines (arriving at right angles to those from London) meant that by the mid-1870s the original station was out of use.

Today the building is used by the Empire and Commonwealth Museum, the old train shed serving as a corporate hospitality function room (*facing page*). The roof is supported on cantilevered wooden beams resting on lines of iron columns; the hammerbeam detail is purely decorative.

Temple Meads was built above ground level to suit the height of the incoming tracks. The whole building is supported on brick vaults, one of which today houses a toddlers' playgroup (*right*). The original fireplace in the boardroom survives (*above right*), as does one of the two coach entrances (*above*).

3

Broad Gauge: Going West

In 1835, when the GWR was still in its infancy, Brunel was also appointed engineer to the Bristol & Gloucester, the Bristol & Exeter and the Cheltenham & GWR Union railways. Within a decade the South Wales Railway was added to the list, as were the companies in Devon and Cornwall and various others in Somerset. The routes of these companies and their later extensions (with the notable exception of the Bristol & Gloucester) would rapidly be subsumed by the GWR and would help define its network.

The Exeter & Bristol progressed rapidly, and the South Devon line was open to Plymouth in 1849. The Cornwall railway had through traffic running to Truro in 1859, but the West Cornwall line only reached Penzance in 1865.

The South Wales route opened in stages, but through trains from England could not run until Chepstow bridge was opened in 1852. That year the railway reached Carmarthen but trains did not arrive at Milford Haven until 1856. Brunel had grand plans for Milford Haven, envisaging ocean liners meeting his railway terminal. They never materialised, but the town did well from the Irish service and flourished in a much more basic way, as a route for fresh fish 'specials' destined for London's Billingsgate Market – three a night.

However much of a success Brunel's broad gauge track may have been for the Great Western, it caused enormous difficulties where it met with lines of standard width. This was such a problem that the government set up a 'Gauge Commission' as early as 1845 to try to resolve the problem. Trials were arranged to prove the systems' relative worth, and Brunel and Gooch emerged with credit. Nevertheless the odds were against one railway bucking the national trend and it was ruled that all new track should be laid to the standard width. Initially 'mixed gauge' lines were introduced, on which trains of either system could run. By 1869 a process of total conversion began, though it was not until 1892 that the last broad gauge train left Paddington for the West.

As well as using a different gauge, Brunel's lines used iron rail with a different profile. Old 'bridge section' rail can be seen used for fencing all over the west of England.

Facing page: Great Western Railway, west of Dawlish.

Bristol & Exeter Railway

In 1837 the Bristol & Exeter Company was formed to take the railway into the south west. Various other companies followed suit, with lines across Somerset, Wiltshire and into Dorset. The common factors were that they all used broad gauge track; they were all supported by the Great Western (which rapidly absorbed them); and each company had Brunel as its engineer.

The original wooden station overall roof (*facing page*) is still in use at Frome on the Somerset, Wiltshire and Dorset line. The stone signal box (*top right*) is at Weston-super-Mare on the main line, and the water tower (*right*) is at Taunton.

This branch line to the harbour at Watchet (later extended to Minehead) closed in 1971 and was reopened by enthusiasts in 1976. Most of the trains are steam, as seen on the coast east of Watchet (*above*). The signal box at Williton (*facing page*) is original.

At Crowcombe there is a display of broad gauge rail, and at Bishops Lydeard a museum.

After Exeter the route planned by the South Devon Railway follows the River Exe, then runs along the sea beneath sand dunes; after Dawlish it clings onto a sea wall below red sandstone cliffs, protected by breakwaters (*facing page*). As well as this level line the route to Plymouth included steep terrain on the south side of Dartmoor.

The company was advised by Brunel that a good solution would be to utilise a new 'atmospheric' technology apparently successfully employed by the Croydon Railway. Motive power was provided by an evacuated tube with a longitudinal slit in which a piston ran attached to the train. The slit was sealed to hold the vacuum with a greased leather flap. A section of tube set in broad gauge track can be seen at the Didcot Railway Centre (*left*).

Work began in 1844 and eight evacuation pumping stations were built, of which one can still be seen at Starcross (*above*).

By 1847 the first section to Teignmouth was opened. The following January the line had been extended to Newton Abbot, but in September that year the experiment was aborted when it proved impossible to maintain the vacuum.

The Earl of Devon made his architect's office in Newton Abbot (*above*) available for Brunel's use.

West Devon

As engineer to the South Devon and the Cornish companies, Brunel advised using timber for the many viaducts through this landscape of deep, sharp valleys. Although wooden construction meant a shorter lifespan, this may not have been short-sighted. If the railways were a success, a replacement would be affordable in thirty years' time; if not, then the speed and cheapness of construction (standard-sized lengths were frequently used) would have been a sensible saving.

In the thickly wooded valley under the main line at Ivybridge, the double row of piers that supported the original timber structure stands next to the present-day arched viaduct (*facing page*). In the yard of the old station nearby are cottages built for the station-master and signalman (*above right*).

At Ashburton, on the now-abandoned branch from Totnes, an altered broad gauge engine shed remains (*right*).

Broadsands

Unusually, Broadsands on the Kingswear branch had an original stone viaduct.

Brunel had hoped to bring the railway into Plymouth's Sutton Docks (where he was consultant), but the Admiralty vetoed the plan. Instead it ran to Millbay Dock, where he was also involved, and remnants of his work can still be seen (*above*).

'Tiny' (*right*) is not directly linked to Brunel, but is the only remaining engine from the broad gauge era. It is housed at the Buckfastleigh Railway Centre and is thought to have been engaged in work at Sutton Dock, where a section of 'mixed gauge' (broad and standard laid together) remains set in cobbles on the quayside (*facing page*).

Royal Albert

The Royal Albert Bridge, crossing the River Tamar from Plymouth to Saltash in Cornwall (*right* and page 72) is considered Brunel's masterpiece of civil engineering. It was the final development of his tubular beam bridges (Cumberland Basin, Chepstow) and utilised the chains made for his then unbuilt Clifton Suspension Bridge.

The Royal Albert opened in May 1859, four months before Brunel's death, completing a through-line from London to Truro. The view on the facing page shows the bridge's silhouette intermingled with that of the adjacent road bridge.

The huge curving tubes of rivetted wrought iron carry the suspended deck in two sections, supported at each bank by stone piers; at the centre of the river the pier is made up of four octagonal cast iron tubes braced together. Essentially the structure is two suspension bridges in line, but because the line is on a curve the forces which would normally be anchored by land-based towers are braced by the overhead tubing.

The bridge was an inventive solution but despite its justifiable fame it was complicated and costly; arguably Brunel's work at Windsor and Balmoral (pages 112, 114) have proved to be more influential in subsequent bridge design.

Cornwall

The main feature of the Cornish and West Cornish railways was their timber viaducts. None survives, but a model of the viaduct at Ponsanooth, on the Falmouth branch (*bottom left*) is on display in 'Steam' museum at Swindon. Today at Ponsanooth some of the stone piers for the timber structure remain (*facing page*). Another example of the old piers standing with the new presents a splendid spectacle from the A38 just west of Liskeard.

More puzzling is the viaduct on the main line at East Taphouse (*bottom right*). There is no evidence of a second row of piers, but the top section of stonework appears to be an addition to support the modern steel truss. What happened to the railway while this work was carried out?

Above: Association with Brunel is often claimed on the slenderest of evidence, but Brunel Quays at Lostwithiel has a fair case: the residential development in the old station yard includes two conversions of original railway workshops.

From Swindon, the other main line of the GWR branched off towards Gloucester and ultimately South Wales.

Although the Railway Acts gave compulsory purchase powers, it was customary to accommodate the concerns of the local gentry. At Kemble a certain Squire Gordon, whose land Brunel needed to cross, exacted a heavy price: an agreement was made whereby the railway line in front of his house was to be hidden in a tunnel and no station built there. Although Kemble was a junction for both Tetbury and Cirencester, it did not have a proper station or appear in its own right on a timetable until 1872, when Gordon died. The water tower (*above left*) may date from the junction at Kemble, not the later station. Today the 'cut and cover' tunnel remains (beyond the Network Rail workers) past the end of the platform of the Brunel-style station (*facing page*).

The contractual problems were equalled by the physical difficulties of dropping through the Cotswold Scarp. The one-and-a-half-mile Sapperton Tunnel gives way to several viaducts on the approach to Stroud; all were wooden and, as there is no hint of any older structures, it has been surmised that the replacement brick arches (as at Frampton Mansell, *below*) were built around the original timber.

Above: A length of so-called 'bridge section' rail beside the main line at Sapperton is in use as a fencepost.

An original GWR contract lives on at St Mary's Mill, a mile west of Chalford, where the line had to cross the access road. Here the mill owner insisted on at least one train a day stopping at the 'halt', a carriage and dining car being provided for an annual employees' outing to London; and the crossing being manned in perpetuity.

The first two conditions have faded away with the 'halt' itself, but the owner of the mill, resisting all offers of automation, still insists on the third. Brian Williams is one of a team of four Network Rail crossing-keepers who man the gates twenty-four hours a day.

In Stroud station yard (interior view, *far right*, and *facing page*) one of the original goods sheds is now in the care of a preservation group.

Right: At Standish, just after the junction with the Bristol & Gloucester (page 82) a cast-iron bridge takes farm traffic over the line.

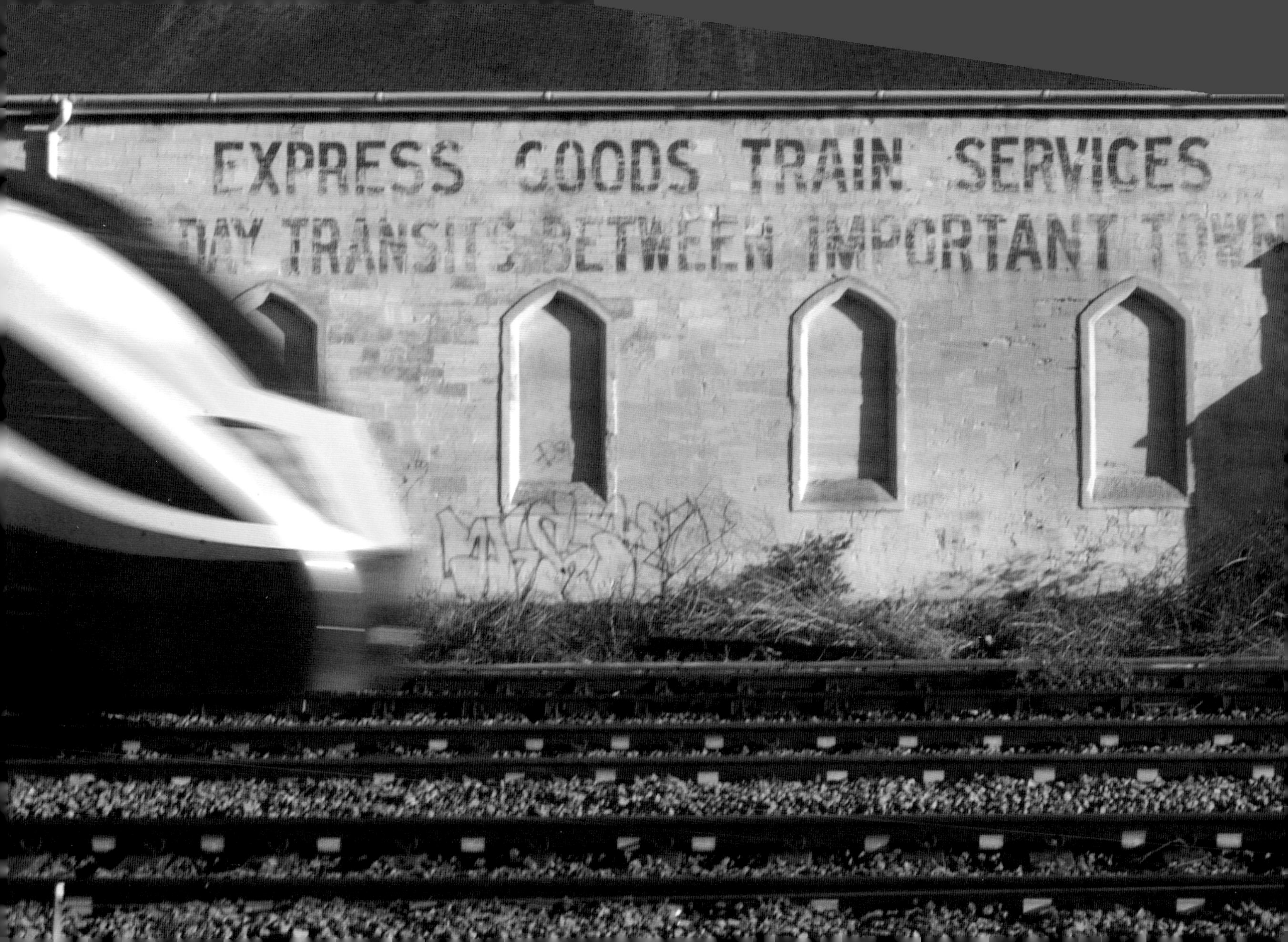
EXPRESS GOODS TRAIN SERVICES

Above: Terry Timperley, Safety Maintenance Officer with Network Rail, indulges in a little role play at Mickleton Tunnel.

In 1845 the contractor of the tunnel works was in dispute with the Oxford, Worcester & Wolverhampton. As the company's engineer, Brunel decided to resolve the issue and marched on Mickleton with a thousand workers from other sites. After a bruising confrontation, the contractor accepted the company's terms and work resumed. This is said to have been the last battle between private armies fought on British soil.

Gloucester (*facing page*) was the focal point for what became known as the 'battle of the gauges'. The Bristol & Gloucester Railway had been built as broad gauge; the Great Western allowed themselves to be outbid for its purchase by the Midland, whose Gloucester to Birmingham line was standard gauge. The inconvenience caused at Gloucester for Bristol-to-Birmingham traffic was a factor in parliament ruling against the broad gauge (despite its excellent performance). For many years 'mixed gauge' track (*above*, at Didcot Railway Centre) was laid that so routes could be worked by both types of locomotives and stock.

Hereford, Ross & Gloucester Railway was opened in 1855 and closed a hundred years later. Today it only features as a line on the map. On the ground its shadowed traces play host to disconnected footpaths, sheep enclosures and lorry parks.

The bridge at Blaisdon (*left*) is typical in its use of rough hewn local 'Ross-red' stone. At Ross-on-Wye a much altered goods shed survives (interior, *above*) as does the original engine shed (*facing page*). The brick arch is evidence of its broad gauge origins.

Ross-on-Wye to Hereford

From Ross to Hereford the line crossed the River Wye four times on timber bridges. The old piers at Backney (*right*) seem too tall to have supported a wooden structure so may be the remains of a subsequent replacement.

Brunel had hoped to take the South Wales line across the Severn at Hock Cliff, a few miles south of Gloucester. In fact, the line left the city on the tracks of the Hereford railway and, from Grange Court, on those of the GWR-adopted Gloucester & Forest Railway. The old Grange Court station sign decorates the exterior of Junction Inn (*above*).

From Bullo the line follows the west bank of the river towards Chepstow, frequently making a picturesque route, as at Gatcombe, where the lines squeeze beneath the river cliffs (*facing page*).

It was always the intention of the South Wales company to purchase the Bullo Pill tramway, which carried valuable coal traffic to the Severn from Cinderford. By 1854, without interruption to this trade, the line – including the thousand-yard-long Haie tunnel – was converted to broad gauge (described by Brunel as 'a tedious and difficult work'). Another important tramway, the Severn and Wye, which reached the river at Lydney, remained independent, but was also converted to the big gauge in 1857.

A piece of broad gauge rail acts as a fence corner above the western portal of Haie Tunnel (*left*); a grassy swathe defines the old route beyond.

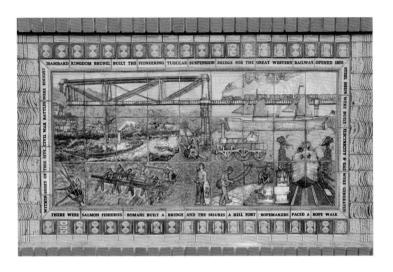

Where the line crossed the Wye into Wales the Admiralty required a high clearance. Brunel designed a hybrid bridge based on a development of the beams he had used at Cumberland Basin, suspending the deck from a tube made of riveted wrought iron. This main section was replaced in 1962 with a welded steel truss beneath the deck, but the cast-iron piers remain (*facing page*).

A piece of the original tube (*above*) is stored at the adjacent engineering yard.

Decorative tile work at the nearby Tesco store (*top right*) gives an impression of the original appearance of the bridge. The cast-iron capital (*right*) is typical of those to be found in many Brunel and Brunel-style stations.

As in the west of England, Brunel used timber extensively for the viaducts in South Wales. This model (*right*) of a section of the bridge at Landore, north of Swansea, is in the care of the City of Swansea Museum.

Close to the modern viaduct at Landore are the unique Llansamlet Arches. Four double flying buttresses (two in view, *above*) braced between the slopes of the deep cutting enabled a steeper gradient, so saving land and labour. Their efficacy is suggested by the fact that in modern times they have been added to with a second arch of masonry and a further weight of soil dumped on top.

G.W.R.
LANDORE VIADU
BRUNEL ENGI
Total Length = 178
Number of Spans 8
Span of Main Truss

Taff Vale Railway

Brunel was first involved with a proposed Merthyr-to-Cardiff railway in 1834 through contacts he made when purchasing iron for the Clifton Bridge. Surprisingly pragmatic, he advised that standard gauge would be appropriate for the difficult terrain, which was so steep that the railway originally included a 1:19 gradient rope-hauled section.

This was near Abercynon, where in 1804 the Penydarren Tramroad was the first line to use a steam locomotive. The engine was built by pioneering Cornishman Richard Trevithick, and a model is at Cyfarthfa Museum (*below*). At Quakers Yard, deep in the tree-choked ravine of the Taff, the old tramroad, (today a tarmac lane) is crossed by Brunel's viaduct (*facing page*). Later widened, the original octagonal piers are closest to the viewpoint.

At Pontypridd, Brunel's Taff Vale line crosses the Rhondda river with wide skew arches (*above*). The shorter spans of subsequent widening, behind the original, are the work of a less ambitious engineer.

Barlow Rail

'Barlow rail' (*above*, in Didcot
Railway Centre) was used on
broad gauge lines and, unlike
bridge section, was laid direct
onto the ballast with no timber. It
was a failure, and all replaced. It is
much rarer to see, but the arched
sections of Clevedon Pier (*facing
page*) are made from it.

The terminus for the South Wales Railway was to be Neyland, on Milford Haven. Renamed 'New Milford' before reverting to its original name, it was later overtaken by the new port of Milford Haven five miles away.

Today the water's edge at Neyland (*facing page*, with the new Pembroke bridge high in the background) is fenced off with old bridge section broad gauge rail. Other examples can be found in the form of fence posts all around the west of England – as in the Swansea buttercup meadow (*right*). At Blaenafon Iron Works, which supplied rail to the world, a length lies unnoticed in the yard (*far right*).

4
BEYOND

When Brunel emerged from a meeting of the Bristol railway committee in 1833 and announced that the new line would be called the Great Western Railway, he did not envisage Bristol as the final destination; nor were Plymouth or Milford Haven to fulfil that role. The shining star on Brunel's western horizon hovered across the Atlantic over New York, with his steam train passengers connecting to steam-powered ocean liners. In late 1835 he voiced this message to the GWR directors, to be met with laughter. Yet nine months later the SS *Great Western* was born, the stern post set up in William Patterson's Bristol dockyard.

Brunel's father had designed steam-powered ships but concurred with the widely held view that no steamship could carry enough coal to cross the Atlantic, the reasoning being that a larger ship would need ever more coal. The son disputed the argument: his logic was that fuel capacity correlated with volume, a cubic measure, whereas fuel use was defined by the surface area pushing through the water, a square measure. Proportions would be significant but, to put it simply, doubling the size of the ship would give eight times the capacity and only use four times as much coal.

In 1837 the SS *Great Western* chased the smaller *Sirius* in a race to be the first steam-powered ship to reach New York. It lost by a day, but made the fastest crossing and vindicated Brunel's theory. Within a year he was planning an even bigger ship, the SS *Great Britain*, with which he was to prove that iron hulls and propeller drive were the future.

And so to what might be considered Brunel's fall. First mooted by him in 1852, the SS *Great Eastern*, designed to trade non-stop round the Cape to India and the East, was so big that even in a period of rapid technical advance nothing larger was built until the *Lusitania* nearly half a century later. This time Patterson and Bristol were abandoned in favour of John Scott-Russell, acknowledged as the premier shipwright of the age, and the banks of the Thames at Millwall. The building of the great ship, which should have been such a triumph, was beset by technical and legal wrangles; in 1859, soon after the ship's maiden voyage, Brunel, ravaged by years of overwork and dispute with Scott-Russell, died aged only 53.

Though the big ships made heavy demands on him, Brunel ran many varied and interesting projects in his later years: more dock work, more bridges and the purchase of an estate for his retirement in Devon, complete with model housing for his workers. He was a member of the committee for the Great Exhibition of 1851 (Paddington Station was enormously influenced by Joseph Paxton's iron and glass structure) and he helped Paxton with the rebuilding at Sydenham (Crystal Palace). Foreign ventures included railways in Ireland and Italy; and when, in 1855, news reached him of the scandalous condition of troops in the Crimea, he collaborated with his brother-in-law, Sir Benjamin Hawes (Under-Secretary at the War Office and reputedly no friend of Florence Nightingale) to design and supply a prefabricated hospital erected at Renkioi.

Facing page: Brunel statue, Neyland, on Milford Haven.

Left: Clifton Woods, whose terraces can be seen through the ship's rigging, provides a splendid vantage point from which to view the *Great Britain* in her dock (*facing page*).

The remarkable features of the *Great Britain* were her riveted iron hull (*left*, and interior view *above*) and her propeller drive, a change from the originally planned use of paddle wheels.

Like the *Great Western*, the *Great Britain* was built in Bristol by William Patterson. The ship was so large that the masonry of the harbour lock had to be dismantled to float her into the Avon.

After her fitting out she enjoyed a successful transatlantic career based in Liverpool. Re-fitted for the Australian trade, she was damaged by fire in the Falkland Isles in 1886, left as a hulk and used for storing sheep's wool. In 1971, her hull and decks intact, she was towed across two oceans to return to the Bristol dock where she was built.

Brunel's second ship occupies a central role in the cultural and historic life of Bristol. Seen by thousands of visitors every year, the SS *Great Britain* has undergone a complete interior refit, with oak bunks in tiny cabins and a splendid dining saloon for corporate hospitality (*top right*). A reproduction of Brunel's original engine has also been installed aboard.

Running through the interiors in a blue plastic tube (*below right*) is an air-conditioning system. Outside, an aluminium grid at the ship's waterline supports watertight glass sheeting on to which a flow of water is pumped. The effect is that the ship appears to be floating in her dock; the practical purpose of the sealed space is to dehumidify the air and conserve the iron hull beneath (*facing page*). *Right:* a contractor spreads the pumped flow of water on to the glass.

The *Great Eastern* was to provide the climax to Brunel's career and probably led to his early demise. In 1859, ravaged by the difficulties of the ship's building and launch, Brunel died, aged 53.

The *Great Eastern* was more than twice as long as the *Great Britain* and with something like five times the displacement. Not surprisingly its construction at Napiers Yard, Millwall, was beset with financial and technical difficulties. Brunel and the ship's builder, Scott-Russell, both men habitually in total control of their projects, were unused to consensus and tended to disagree in the search for solutions to problems. The ship progressed slowly.

Rather like the Thames Tunnel, the building of the *Great Eastern* became a public spectacle. When launch day finally arrived in December 1858, the riverside was packed, but the public were to be disappointed. Because of her length, the ship had to be sent sideways into the Thames and the hoped-for free slide into the water never happened. After months of

delay she was finally launched with the help of hydraulic presses.

Excavation at Napiers Yard (*bottom right*) has revealed the piling and timbers on which the great ship was built and launched. Her size made her ahead of her time in the market for passenger travel, but her huge hold was perfect for the task of laying transatlantic telephone cable. The cable came ashore at Porthcurno

in Cornwall, where today a museum displays exhibits relating to Brunel and the Great Eastern, including a section of the cable (*top right*) and a working model of the ship (*above*).

Facing page: A print of the ship in her heyday is proudly displayed by Ray Harding, museum curator at Milford Haven. The redundant ship had been left there in 1880 and the dock built around her.

On the proving trip of the *Great Eastern* in 1859 a funnel blew off in a water pressure explosion, killing a crewman. After serving for many years as a filter in a Weymouth waterworks, the funnel (*above*) is now displayed in the Great Britain museum.

At New Ferry, Birkenhead, close to the spot on the Mersey where she was broken up, is the 'Great Eastern' pub, reputed to have been built for the refreshment of the workers. Publican Wayne Williams stands in front of the bar, which originally came from the captain's saloon (*centre*). Panelling from the ship (*facing page*) has been transferred from the pub to the Porthcurno museum (page 106).

One of the *Great Eastern*'s anchors stands on the quayside at Concarneau in Brittany (*right*), and is said to have been dragged there after being snagged by a trawler. The mizzen mast, one of seven but the only one made of wood (it carried magnetically-sensitive navigation equipment) serves as the flagpole for Liverpool Football Club's ground at Anfield (*top right*).

When the railway arrived at Briton Ferry, Brunel's help was sought for dock improvement. His first move was to divert the river to make the harbour less prone to silting up. Then, within outer piers (*facing page*) he created an inner harbour using a single lock gate of the semi-buoyant type he had tried at Bristol (*above*). Most of this gate was cut away by a demolition team in the 1980s, but the bottom section remains.

An engine house, now semi-derelict (*top right*), pumped water up into a tower (*right*) to provide hydraulic pressure to drive the gate. It is hoped to restore the tower.

The bridges at Windsor (*previous pages* and *left*) and Balmoral (*right*) both lead to favourite royal palaces. They also have a more important common bond: unlike the more famous Royal Albert Bridge (pages 72-5), they both point directly but in different ways towards the development of modern bridge design.

With the triple 'bow and string' construction of the bridge at Windsor carrying the branch railway from Slough (*left*, and page 112), Brunel appears to anticipate the welded steel bridges which became common much later in the century.

Brunel's friendship with Prince Albert led to a commission to bridge the Dee at Balmoral.

The ironwork, by Brunel's friend Rowland Brotherhood of Chippenham, was equally forward-looking, its riveted lattice work conferring strength with lightness.

Queen Victoria, with her liking for Gothic detail, was not amused by the bridge's elegant simplicity. Perhaps the Queen, who would anyway have been sniffy about 'trade', may have got a whiff of Brotherhood's background: before he became an iron-master he was a cesspit digger. (Brunel found his skills in soil mechanics an asset when building difficult embankments.)

Brunel – Country Squire

Brunel was such a busy man it is hard to imagine him in a domestic setting. In 1836 he married Mary Horsley, and their home at 18 Duke Street, Westminster (now demolished) became the centre of an active social life which brought them into contact with many of the distinguished figures of the day.

While working in south Devon, Brunel was much taken with the area and bought an estate at Watcombe above Babbacombe Bay, intending to move there in due course with his family. He drew up a plan for a mansion on the hilltop site overlooking the sea (which the existing 'Brunel Manor', *above*, may not reflect) and spent time setting out woodlands on the slopes. The estate is being researched and restored (pebble-edged paths, *right*).

Aspiring to be a model employer, Brunel also designed workers' cottages (*facing page*) and a set of larger houses for professional staff at nearby Paignton (*far right*).

As well as his large-scale projects at Paddington, Millwall and Rotherhithe, Brunel built a suspension bridge at Charing Cross, which was later replaced by Hungerford railway bridge. The red-brick abutments (*below*) are part of Brunel's original suspension towers. A prominent statue of Brunel stands on the Embankment nearby (*left*).

Facing page: Through his friendship with Joseph Paxton, Brunel was involved with the removal of the Great Exhibition from Hyde Park to Sydenham. Paxton wanted extensive 'water features' in the new location, and was uncertain of designs he had obtained for two water towers. Brunel took over the job, and the towers became a feature of the relocated 'Crystal Palace'. The low base of the demolished 'south tower' remains (behind the museum) with original water pipes and valves.

5
BRUNEL: A PLACE IN HISTORY

In 2003 Brunel was placed second, by popular vote, in a television programme which aimed to establish the 'greatest ever Briton'. What more can be said about a man whose achievements have placed him in the lists with Shakespeare, Churchill and Elizabeth I?

Of Brunel's peer group, the man most obviously his equal was Robert Stephenson. Also the son of a famous father, Stephenson too was involved in mechanical as well as structural engineering and was prepared to push at the boundaries of technology to achieve his goals. It has been suggested that Stephenson and Brunel were deadly rivals; in fact, despite each advocating a different gauge of railway, they were close and supportive friends. When Stephenson needed help in raising the Britannia Bridge over the Menai, Brunel was there to help. When the fledgling Great Western Railway needed locomotives, Stephenson supplied them. These two men were giants in the golden age of railway expansion. Both realised the importance of maintaining London households: enmeshed in establishment circles, they were ideally placed to lobby government, and did so effectively to further their clients' schemes.

Which leads us to consider a great man from a previous generation. By the 1830s Thomas Telford was finally giving up his nomadic lifestyle to join the establishment, taking up an appointment as advisor to the Treasury. His career, though studded with structures considered at the time to be wonders of the world, was long and steady, a careful consolidation of technological experiment which made him a master of the use of iron. Brunel also had the inventiveness and daring to come up with seemingly impossible schemes and the charisma to persuade his backers to risk their investment. But in comparison with the methodical Scot, the physically fearless and dramatically persuasive Brunel was an unpredictable shooting star.

Let us not forget Brunel's own father: a development of his major invention, the 'tunnelling shield', is still in use. Must we declare a winner? These men all made massive contributions to the Industrial Revolution. Telford left a legacy of working practice which is still acknowledged today; Stephenson's concept of the strength of a riveted conduit, so large that a train could pass through it, led to modern box-girder designs.

So how should Brunel be remembered? Historians and professionals will cite his railway, his bridges and his ships (iron-built, steam-powered, propeller-driven), all of which support his colossal reputation. The disenchantment of investors who lost money in his schemes will be consigned to historical footnotes (a reminder being the public house opposite Bristol's Temple Meads, named 'The Reckless Engineer').

Facing page: The famous photo of Brunel by Robert Howlett, framed with nautical chains at Millwall.

However, it might not be the engineering works that the great British public have in mind when they sing Brunel's praises: it is possible that Robert Howlett's compelling photographic portrait (page 120, framed amongst nautical chains at Millwall, the site of the *Great Eastern* launch) is a more familiar reference. This iconic image, universally known from countless book covers and magazine articles, contains many pointers that help us understand the man: the cigar hints at his flamboyant personality, the top hat tells us of his professional status, the chains reveal the nature and scale of his achievements, while the level gaze to a distant horizon may remind us of the world view into which he placed his many visionary projects.

So for many reasons Brunel is the man we remember best, and it seems certain that flags will continue to fly over the achievements of this great Victorian.

Brunel is buried in the family vault in Kensal Green cemetery. He is also commemorated in a stained glass window in Westminster Abbey.

Index to Photographs

At Gatehampton bridge the railway leaves the Thames, heading west
towards the Vale of the White Horse.

BIBLIOGRAPHY

Brindle, S., *Brunel: The Man who Built the World* (Weidenfeld & Nicholson)

Bristol 200, *Audit for South West* (Culture South West)

Buchanan, A., *The Life and Times of Isambard Kingdom Brunel*
    (Hambledon & London)

Buchanan, R.A. and Williams, M., *Brunel's Bristol* (Redcliffe)

Cossons, N., and Trinder, B., *The Iron Bridge* (Phillimore)

Cragg, R., *Civil Engineering Heritage – Wales and West Central England* (Thomas
    Telford Publishing)

Falconer, J., *What's Left of Brunel* (Dial House)

MacDermot, E.T., *History of the Great Western Railway*, Vol. 1 (Ian Allan)

Pudney, J., *Brunel and his World* (Thames & Hudson)

Rolt, L.T.C., *Isambard Kingdom Brunel* (Longmans/Penguin)

Vaughan, A., *Isambard Kingdom Brunel: Engineering Knight Errant*
    (John Murray)

Hanging baskets at Kemble station.

ACKNOWLEDGEMENTS

Thanks to the following organisations for their help:

Bristol Industrial Museum
Bristol Harbour
Bristol 200
Buffer Bears Limited
Brunel Engine House, Rotherhithe
Buckfastleigh Railway Centre
Chippenham Museum
Corsham Heritage Centre
City & County of Swansea Museum
Crystal Palace Museum

Cyfarthfa Museum
Didcot Railway Centre
Empire and Commonwealth Museum,
    Bristol
Firefly Trust
Liverpool Football Club
Museum of London
National Railway Museum, York
Network Rail
Newton Abbot Museum

Porthcurno Telegraph Museum
SS *Great Britain*
'Steam', Museum of the Great Western
    Railway, Swindon
Stroud Preservation Trust
West Somerset Railway
World of Boats, Eyemouth

Thanks also to those whose images appear in the book, and to the following:

Steven Brindle
Professor Angus Buchanan
Shane Casey
Challoner Chute
Eileen Collins
Sir Neil Cossons
Hugh James

Andy King
Anne Mackintosh
Ivor Martin
Paul Manning
Ben Morris
Stephen Morris
Nigel Overton

Bryan Osborne
Sarah Roberts
Marc Steadman
Mike Stone
Martin Yallop
Ridley Youngman

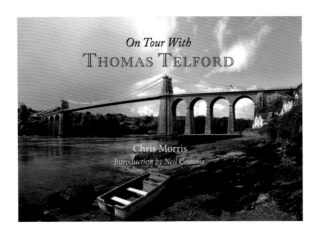